Math Without
A Migraine

Math Without
A Migraine

Shauna M. Johnson, M.Ed., N.B.C.T

ISBN: 148008798X
ISBN 13: 9781480087989
Library of Congress Control Number: 2015918819
CreateSpace Independent Publishing Platform
North Charleston, South Carolina

"Learning math concepts boring? Not in Shauna Johnson's classroom. As an educator, I was fascinated as I watched Shauna transform her class into a stage where she creatively brought math concepts to life. Her stories poured out with enthusiasm, her students were captivated, and more importantly they were learning and retaining valuable information. Shauna's stories seem to tap into her student's world, simplifying difficult concepts and making them memorable for years to come."
Linda Barrett
Fellow Educator

"I never got the mean in mean, median, mode and range until Mrs. Johnson explained it in a fun way. She would tell me that everyone thought the oldest child (me) in the family was mean. She would explain everyone just misunderstood the mean, but he was just average--she emphasized the just. So to find the real mean you had to find the average! It has helped me with mean to this day!"
Carolyn Rouse
Former Student

"Mrs. Johnson is a great story teller. I witnessed her teach many young math students with her imaginative "down on the farm" story of median, mode, mean and average. The children are captivated by the entertaining tale. Learning math is a journey and Mrs. Johnson makes sure that the children have a great time along the way."
Cindy Saylor
IA, FCPS

Acknowledgments

Books are born every day, invariably expressing the thoughts, knowledge, spirit and inspiration of the author. In order for these endeavors to come to fruition, it often requires the help, guidance, and tutelage of many supportive friends, family, and co-workers.

For this, I have many to thank for their patience, support, prodding, and encouragement to make this dream a reality.

To Linda Barrett, who constantly told me I needed to write these stories down and share them, you started this journey for me and planted the idea in my mind.

To Cary and Catherine Guidici, who took my early drafts and brought my thoughts more vividly to life, you made my work fun.

To my children, Brandon, Nichole, and Alyssa, who encouraged me and made me believe I could do this, I owe you the world. And, especially to my daughter, Alyssa Johnson, who had to sit through my lessons as a fourth grader, and then as a high school senior, drew several of the pictures for the book, you definitely pulled double duty.

And to my husband, who has always held me up as I reached for the stars, you made this dream a reality.

Foreword

So many years as a teacher, so few things that haven't changed. Some change was good; some, I must admit, was less satisfying.

Twenty-five years ago there seemed to be more joy in going to work each day. We looked forward to crowds of eager, happy students entering the classroom to explore and create together. We undertook seasonal art projects for the back bulletin board, and holiday musicals. We might stop in the middle of a lesson and linger on a student's especially thought provoking question.

Then the focus became more on test scores, and less on nurturing our students' natural curiosity. We began turning into an educational society of "surface dwellers." Students were prepared to answer 'who," "what," "when," and "where," and other questions designed to be easily graded on a multiple choice test. We never had time to ask "why" and "how." Students' joy in the journey and love of discovery lost out to a new ideal: their ability to spew facts and figures.

But there's still hope, and you feel it! When we close our classroom doors, we open up a world of possibilities. Of course the curriculum must be taught. Yes, student achievement remains our top priority. But nobody can or should take away our freedom to choose HOW we teach, to bring back that light in our students' eyes during one of those unforgettable "Aha!" moments.

How can we instill a love of learning—which is still, after all, our ultimate goal? We can't teach a child everything he or she needs to become successful in life. We can teach them the power of asking "why" or "what if," to take risks, and celebrate making mistakes as the fastest way to further learning.

For a child, the importance of a teacher in his or her life is second only to a parent. That's because we take time to "show me I can do it." That's why we're teachers, and love it enough to get up at the crack of dawn and think, create, cajole, purchase, grade, cut, and paste 'til dark.

My students' natural tendency to relate, to connect, to imagine, and to feel between the lines of our jam-packed curriculum continues to light a fire inside me, which keeps burning strong even after so many years. WHAT I taught students was never as important as HOW, when helping yet another group feel excited about a new math lesson or the story of the Revolutionary War.

This book gives me an inspiring opportunity to share my successes with you. I hope it will help you see a new sparkle in your students' eyes every time you get ready for math, science or reading. We both want to help students love learning, and we know they can succeed at it, as we know just how great it is to be a teacher.

On behalf of all your students—past, present and future—and their families, thank you for taking time to consider my little book. And an even bigger thank-you for your lifetime of commitment to the people who always matter the most to any teacher: our wonderful students.

Table of Content

Introduction

Contemplate the depth and complexity of a teacher's responsibilities. They serve as instructors, advisers, guides, trainers, tutors, lecturers, coaches, and facilitators.

In an ever changing world of students, standards, and curriculum, amid mounting social and political pressure and expectations, teachers must shoulder the burden of our future.

For 25 years as a teacher, my greatest challenge has been to work smarter rather than harder. Now, enhancing educational best practices with enduring research on how the brain learns makes it possible to reach more students, more effectively, in less time than ever.

We understand that all learning is brain-based. So every teacher must be familiar with some basic features of how the brain works in the learning environment.

Learning requires sustained attention, but fatigue begins affecting the brain's neural systems within three to five minutes of such activity.

Consider the example of a violin string. When a single note is plucked, an initially strong sound begins to fade. However if the string is repeatedly plucked, the tone continues strong and clear.

Any student listening to a long lecture absorbs some information, but their attention quickly begins to fade. By repeatedly "plucking" his or her neural systems by alternating between facts and concepts or context, we help a student pay close attention for longer intervals.

For example, when a child hears you say, "Thomas Jefferson was our third president," one neural system is used. When a related concept such as "But he was much more than just a president" is communicated a different, interconnected neural set is activated. Introduce a short vignette from Thomas Jefferson's life, and a third type of neural system will be fired.

Any student who receives information in this way will remain attentive for a longer period of time, because more of his or her neural systems are being engaged.

Dr. Bruce D. Perry, M.D., Ph.D., an internationally recognized authority on brain development has written, "The most effective presentation must move back and forth through these interrelated neural systems, weaving them together. These areas are interconnected under usual circumstances, like a complete 'workout' in the gym where we rotate from one station to another. Similarly, in teaching, it is most effective to work one neural area and then move on to another. Engage your students with a story to provide the context. Make sure this vignette can touch the emotional parts of their brains. This will activate and prepare the cognitive parts of the brain for storing information. Information is easiest to digest when there is emotional 'seasoning'-humor, empathy, sadness, and fear all make 'dry' facts easier to swallow. Give a fact or two; link these facts into related concepts. Move back to the narrative to help them make the connection between this concept and the story. Go back to another fact. Reinforce the concepts. Reconnect with the original story. In and out, bob and weave, among facts, concepts, and narrative. Human beings are storytelling primates. We are curious, and we love to learn. The challenge for each teacher is to find ways to engage the child and take advantage of the novelty-seeking property of the human brain to facilitate learning."

Brain based learning can be considered a great unifier. Students from a wide variety of cultural, linguistic, political, socio-economic, and family backgrounds all essentially learn in the same way. Maximizing brain function makes it easier for all students to learn.

How can a knowledge of brain based learning help you teach mathematical principles such as multi-digit multiplication?

Here's the traditional approach to teaching math: the teacher introduces a concept or skill, students practice, often as a group. You check for understanding, and then have students independently complete a set of similar math problems.

But brain-based research tells us this type of teaching and learning is much less stimulating, which greatly lowers the amount of attention the brain makes available to students. Students understand less and remember less.

The math activities described here are designed to add "bobbing and weaving" to your lessons and improve students' focus and attention. Each features an emotionally engaging, somewhat familiar story, mathematical facts, and concepts your students can use as building blocks for future understanding and long-term classroom success.

notes

CHAPTER 1

Number Sense: Do These Numbers Make Sense?

Playfulness, socialization, curiosity, and exploration. They drive our innate urge to learn from the time we first learn to walk and talk.

"Mommy, what makes that traffic light change colors?"

"Daddy, how do you tie your shoes so good?"

"How do you draw a kitty?"

"Oww that's hot!"

"Eww, that tastes yucky. I'll never eat that again!"

"How come Jane has so many blocks? How can I get more?"

"Ha ha, look at how Spot runs in circles! That's funny!"

Older children learn differently, but they never completely lose their passion for that survival skill called learning. So how can teachers bring students' curiosity, socialization, playfulness, and exploration back into the classrooms?

Teaching, in an important way, is brain science. Discovering new information on how we learn inspired me to write this book. Each chapter features short vignettes and wacky versions of familiar fairy tales that have captivated my students for years.

Even the most dedicated teachers can get discouraged by the feeling that many of their students think a carefully planned and executed math lesson is like a visit to the doctor. Ready to make your tired old lessons disappear? Then let me introduce you to a new trick for your magician's bag.

These fractured versions of familiar stories and collections of oddball characters will do some heavy lifting of abstract concepts. They'll help you reconnect with the love of learning that still bounces around in your students' brains, by adding more playfulness and curiosity to each lesson. Your students will love this exciting new perspective on essential math concepts and formerly "dry" concepts and skills.

There's no guesswork here. I've seen enthusiasm build in students as facts, context, and concept are mixed for a more productive learning experience. With neurons firing at optimal intervals, learning is always at its peak.

Ready, get set . . . learn!

The Place Value Families

Once upon a time, in a land far, far away, a small family decided in their weekly family meeting to escape the hustle and bustle of the big city. They packed up their prized possessions—hair curlers, fishing nets, stuffed animals--and made plans to move into a perfect little yellow house under a big blue sky out in the country.

This family's dad had thinning hair, the mom lots of silk scarves, and their little girl a fascination with Mom's high heels. Every night Dad came home from work, his imitation leather wallet filled with $100.00 to support the family.

Every day Mom borrowed some shoes back from her daughter and went shopping for food, clothes and maybe a goldfish. An excellent shopper, she knew how to spend exactly $10.00 every day. And every day the little girl worked so hard folding laundry and doing other chores, her parents gave her $1.00.

They really liked each other, and having a good sense of humor gave each other funny nicknames. Dad was "Hundred", Mom was "Ten", and-you guessed it-the little girl was "One." At least that's what they always called each other.

Illustrate a place value chart or house and color code each place value so they correspond with the vignette's three characters. Briefly discuss the first three places of place value.

HUNDREDS   TENS   ONES

Life was just peachy for this family, until one day when another house just like theirs appeared next door (in a made-up story like this, houses will appear overnight). You can imagine how upsetting this was for them. But in an emergency family meeting they decided to become good friends with the new family.

"Ten" made chocolate chip cookies and arranged them on a plate. "One" carried the plate. "Hundred" tagged along to say hi. The moving truck was just driving away as little "One" carefully balanced the plate and rang the doorbell.

To her surprise, a little girl who looked just like her opened the door! "Hi! My name is One!" exclaimed the first little girl, almost dropping the plate in surprise. "Wow--me, too!" shouted the second little girl from inside her doorway.

And guess what? This new family's dad also brought home $100.00 every day in an imitation leather wallet, and the mom took care of her family by spending exactly $10.00 every day, and their daughter worked hard enough on her chores to earn $1.00 every day. You'll never guess this part of the story-- these neighbors even used the same nicknames for each other!

That was a problem! The first family had been the only family around, so they'd never had to use their last name.

In fact, they'd even forgotten what it was. Of course, both families couldn't be known by the same nicknames--Hundred, Ten, and One. So, they held a joint family meeting right there in the living room and came up with a great idea.

Since--naturally--the second family didn't have a last name either, they agreed on a perfect new name. Because they had traveled a thousand miles to their new home, they would be known as the Thousand family.

Point out the repeating patterns in place value through the hundred thousands place.

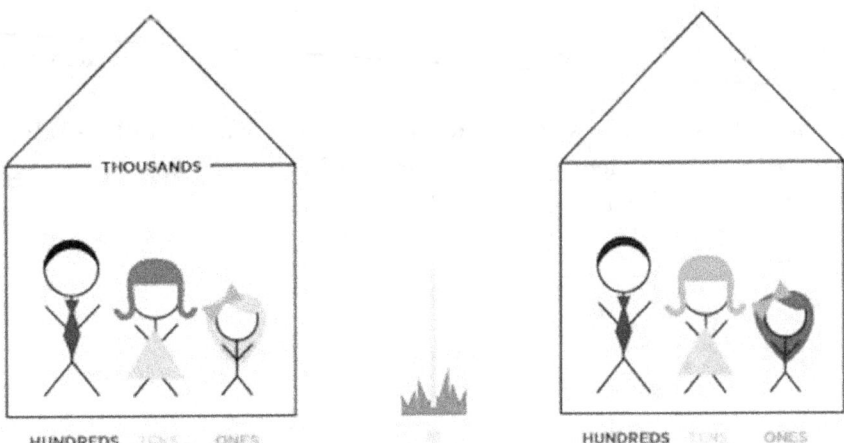

And for a while, everything was perfect. The families went on picnics together and had barbeques in their shared backyards. The girls frolicked in the meadow, picking flowers (**all great things to act out and have fun with.**) Life was good.

Until that fateful day when another new house went up right next to the Thousand's home. And you'll never guess who moved in. Yep, the newest family's dad had thinning hair, the mom piles of silk scarves, and their little girl was exactly the same age as the other two little girls. Do you know what they called each other? That's right. Hundred, Ten, and One.

This family was nice enough. And the first two families wanted to get along with them because the new-comers were really, really rich. So rich, they'd adopted a last name to let people know just how rich they were. Introducing….the Millions!

Stop and illustrate place value through the hundred millions place. Explain how it relates to the concept of place value.

Needless to say, after getting to know each other, the three families became very close friends and lived happily ever after!

Resuming the math lesson, make a connection between the story and place value. Students are now ready to assimilate their new knowledge with the familiar. In this way they increase their understanding and ability to recall this new information when called upon to do so.

Adding Decimals to the Story

When expanding the concept of place value to include decimals, upper grade teachers may want to add details that illustrate how decimals fit into the picture of place value, and especially why there is no "oneths" in decimals.

The three families **(more for higher grades)** lived together for years in peace and harmony. They borrowed sugar, tools and stuffed animals from each other. But then things began to change again. A two-lane, very busy road went in alongside their houses, and across it a new apartment building with a sign that read, "No Children Allowed!" How rude was that?! This meant only grandmas and grandpas could live over there!

Create an illustration to help students understand the "big picture" of place value. Review facts about place value through the hundred millions place, bringing out the repeated patterns.

These grandmas and grandpas were very sad their own grandchildren couldn't come and visit! Without families, all they could do was look at family photos and do crossword puzzles. And, the three girls couldn't go visit. At the busy road, they had to STOP! **(at the decimal point.)**

The girls knew the grandparents were sad; even the flowers in the window boxes looked droopy. So they decided to do something about it. Not being able to cross the busy street, they decided to holler over the busy road to the people.

So One, One and One took some real deep breaths, stepped up to the edge of the busy road, and waved for the grandmas to come closer to the other side of the road. Then the girls yelled over, "Hi! Our names are 'One'! What are your names?"

"Well," one grandma yelled back, "we all used to be named 'Tens', but being so old we call ourselves 'Tenths' now." She sighed. "We feel like we're just part of what we used to be."

"Oh," all the girls sighed back at them. Then the grandpas, curious about all this sighing, joined their wives on the side of the road.

When the girls saw them they hollered again, "Hi, grandpas, what are your names?"

"Well," hollered one grandpa, "we used to be called 'Hundreds.' But we're getting older too and feel like we're just part of what we used to be. So now we call ourselves 'Hundredths.'"

Soon the three girls and the grandmas and grandpas were best friends. And even though they couldn't be together on the same side of the road, they spent many hours yelling back and forth to each other. They sent Christmas cards and birthday cards and the grandmas and grandpas were happy.

Which is why there are no "oneths" in decimals. Because there are no children allowed on that side of the road.

Explain that a "oneth" really means one over one, or a whole.

During these stories, students' minds constantly move between facts, concepts, and context. Their brains pay closer attention to the entire lesson, process information more efficiently and store it for a higher rate of recall. Students have made an emotional--hopefully humorous--connection with place value as their brains get the curiosity, playfulness and exploration they need. And a teacher's ability to immerse students in the concepts of place value will yield significant rewards.

Rounding and the Tower of Terror

The concept of rounding is taught consistently throughout elementary school. Many students find it difficult to understand its multi-step thought process, especially when deciding if a number should be rounded up or down.

Using this short, personal vignette, students get to pretend they're on a scary ride, tapping into the brain's emotional connection. We also get to "bob and weave" between facts, context, and concepts of rounding.

"Have any of you truly been scared out of your wits before?" I first ask students. We share what scared us, and how it felt. This quickly connects the brain to one of our strongest emotions, fear.

A teacher I know once got really, really scared! One summer her family went to Disney World in Florida, and her teenage son asked her to go along on a ride called the Tower of Terror.

Establish a connection to what you're talking about by mentioning a similar ride in a nearby theme park.

If you don't know this kind of ride, it goes up and down a really tall tower. First it carries you up several floors in a double row car. It puts the two rows of seats inside a kind of elevator.

Just as you wonder who had this crazy idea, you're strapped in and the elevator doors close. You wonder if it's too soon to try a little scream. But the people with you are concentrating for what seems like a long time as the ride lifts you higher and higher, and just as you think maybe you can get out...BOOM, the car drops!

With the windows open it's easy to see just how high up you are. Then you shoot back up the tower, unsure if that was you screaming or somebody else. The elevator keeps jumping up or down unexpectedly, until it finally reaches the bottom of the tower and you can breathe again. Yikes!

This teacher's usually pretty brave when it comes to roller coasters and rides. But this ride looked real scary, so she was glad her son came along. And the whole time she was on the ride, she was squeezing his hand so tight he thought she would break some bones. But, she was NOT going to let go!

Rounding numbers is a lot like that Tower of Terror ride, but I won't strap you in. When doing a rounding problem, you're supposed to round to the nearest place value. Whatever place value being asked, becomes "her" on the Tower of Terror--if you need to round to the nearest tens, that tens digit is "her". Or if you're supposed to round to the nearest hundred thousands, that hundred thousands digit is "her".

Stop and briefly review place value. Write numbers on the board and ask students to name the digit in the place value you call out. This is a quick assessment for you to know if students understand which digit is "her."

This teacher's son is younger, so in place value, he always stands for the number one place value less than "her." That's the number you hang onto and NEVER let go. That's the place value you ALWAYS look to when rounding.

Going back to the random numbers you've put on the board, have students again identify the digit that is in a certain place value, the one that represents "her." After locating that digit, have them look for the digit that represents the son, the digit in the place value one place to the right of "her."

Here's what's even better with rounding than on the Tower of Terror. In rounding, you KNOW if you're going up or down. But in the Tower of Terror, you never knew which direction you'd be going.

There's a helpful saying used to remember which direction to round. "5 and above, give it a shove, 4 and below, let it go." So if the digit that represents the son is a 5 or higher, the "ride" is going up, and the number representing "her" is going up by one digit. If the digit for the son is 4 or lower, the ride is going down, while the "her" number stays the same.

Incorporating movement by having students stand up or sit down depending on whether they would round up or down, also helps keep students' attention on learning.

Creating a class room in which playfulness, curiosity, socialization, and exploration are always active will help students learn. The more you keep their brains actively engaged in the learning process, the more effectively they will understand the subject matter. Constantly "bobbing and weaving" between fact, concept, and context ensures that your students will always love to learn!

notes

CHAPTER 2

Operations: But Not Like a Doctor

Our brains, natural pattern-seeking processors, have a phenomenal ability to organize the most random and seemingly chaotic information into a predictable order. This helps us survive and flourish amid the random, sometimes chaotic moments which, tied together, fill our days. If we make it easier for our brains to use their pattern-making skills, they reward us with mental clarity and insights into the more subtle meaning of such murky moments.

Patterns play an especially important role in mathematics. As I often remind students, math is all about patterns. Once you find the pattern, math is easy.

Long Division: Goldilocks and the Three Bears

Once upon a time (as all good stories begin) a curious little girl named Goldilocks lived in a beautiful forest. One day she was taking a walk, chasing butterflies and thinking about a cute boy who'd just moved into the neighborhood, when suddenly she became lost and afraid.

Wandering in circles through the woods trying to find her way home, she discovered a cave – just a hole in the hillside, half hidden behind a hedge. Being curious and friendly, she carefully parted the hedge and called into the cave, "Hi, anybody home?"

She had stumbled across the home of three bears: Papa Bear, Mama Bear, and little Baby Bear. Papa Bear, hearing the little girl's voice, lumbered out to investigate.

"What's the matter, little girl?" asked Papa Bear.

"I'm lost and hungry. Do you have any porridge I could have?"

Announce a break in the action, then write out a long division problem that includes a three-digit dividend and a one-digit divisor. Point out that in this long division story, Goldilocks is the divisor. She approaches the "cave", or long division sign, and talks to Papa Bear first. Papa Bear represents the number in the hundreds place. Briefly refer to the familiar place value story from chapter 1 to make a stronger connection.

Papa Bear's answer to Goldilocks' first question depends on which digits represent the two characters and if Papa Bear is greater than Goldilocks. For example, if Goldilocks is represented by the number 5 and Papa Bear by 7, Papa Bear must answer "yes". If Goldilocks is a 9 and Papa Bear a 4, Papa Bear's answer would be "We need to check with Mama Bear." In this latter case, Goldilocks would next ask Mama Bear and Papa Bear together. In long division terms, students must look at the hundreds and tens- place digits together as one number. In the example above, with Goldilocks being a 9 and Papa Bear a 4, we would add 2 for Mama Bear. Students should get used to looking at both Papa Bear (4), and Mama Bear (2), then come up with the number (42) that Goldilocks would be "talking" to.

If Papa Bear is larger than Goldilocks, on the other hand, students should proceed through the steps of long division using the acronym DMSCB (divide, multiply, subtract, check, bring down) so they can work through the steps sequentially. To make memorizing this acronym easier, help them think up silly phrases like "Does McDonald's Serve Cheese Burgers?" or "Does My Sister Chase Boys?"

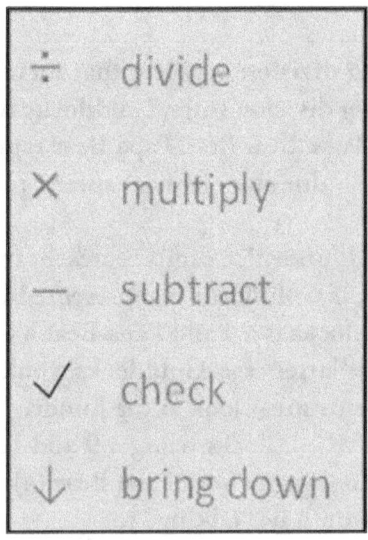

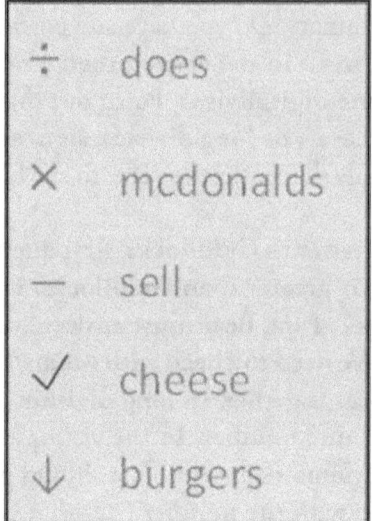

When students have completed the first step in the long division process, and brought down the tens place digit, resume the story.

Goldilocks loved the porridge so much, she went back to the cave and asked Mama Bear, "Do you have any porridge I could have?"

Repeat the pattern you used with Papa Bear, this time with Mama Bear.

After finishing her second steaming bowlful of porridge Goldilocks was still hungry, so she went back to the cave and now asked Baby Bear, "Do you have any porridge for me?"

Repeat the pattern a third time for Baby Bear.

After her third bowl of porridge, the three bears walked Goldilocks back to her forest home. Before they left her, Goldilocks promised to repay their kindness with free babysitting for a month. And they all lived happily ever after.

Several patterns are used in this story to help brains process. Repeating steps for each place value and the use of Papa, Mama, and Baby as in the place value vignette, establishes patterns that the brain will find easy to recognize. This makes it easier to recall and make sense of the steps used to successfully answer long division problems.

Multi-Digit Multiplication: Jack and the Beanstalk, Sort of
As with long division, our multi-step multiplication process is designed to establish patterns and processes that the brain needs for optimal performance.

Once there was a boy named Jack, who tried his best to stay out of trouble. Every day he'd get up early and help his mother around the farm by milking the cows, feeding the chickens, and sweeping their cottage. Although he knew these tasks were important, his secret dream was much grander: to become rich!

One morning Jack's mother told him to go to town and sell their cow, because they needed money to buy seeds for their garden. Jack felt quite sad. Bessie the cow was his best friend. But he did what his mother told him without arguing, so he would be a good son.

On his way into town, Jack ran into someone he'd never met before. This stranger wore a shiny blue suit, and had lots of yellow hair piled high on his head. Pointing at Jack, he shouted, "Would you like to have lots of money?"

"Of course!" exclaimed Jack. "Then my mom and I wouldn't have to work so hard."

"Then let me give you these magic beans in exchange for your cow!" exclaimed the excited stranger. Jack really wanted to be rich, so it only took him a moment to agree to this unusual offer. He handed over his beloved Bessie, in exchange for three magic beans the stranger dropped into his hand before walking away, waving his arms in the air.

Give each student 3 "magic" beans.

Jack ran home to tell his mother about the magic beans! Of course, she was not nearly as excited as the stranger had been. Jack had forgotten to ask what made the beans magic, or how to use them. Now he had to agree that he'd probably been tricked. So he threw the beans onto their kitchen table, thinking that his mother might make them into soup. Very thin soup.

That night, Jack was at the kitchen table doing his homework when something magical happened. He had to complete some really hard math problems that his teacher had called "multi-step multiplication". Jack could hardly even say those words, much less solve the problems.

But he tried his very best to remember what his teacher had told the class, without much success.

Students should work along with Jack on their white boards or papers as they see what happens next in the story.

"The first thing, my teacher told me, was to write down the problem going vertically. That part's easy," Jack said out loud to himself.

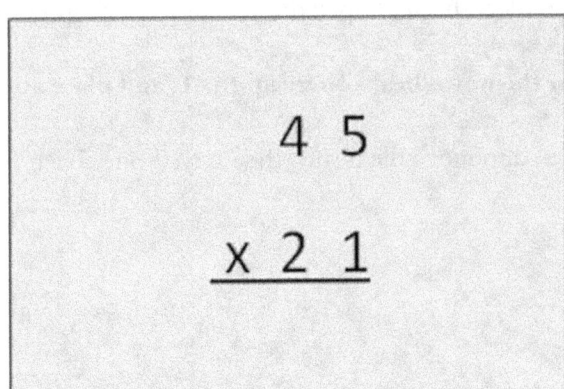

$$\begin{array}{r} 4\ 5 \\ \times\ 2\ 1 \\ \hline \end{array}$$

Have students write a 2-digit by 2-digit multiplication problem vertically on their white boards or papers. For example, 45x21 above.

But wow - as soon as Jack wrote the problem down vertically, one magic bean jumped from the middle of the table onto his paper and covered up the "2" in 21!

Have students also make their beans jump over so it covers the 2 in the problem.

Jack, startled, tried to pull the bean off his paper--but it wouldn't budge! "Hey! I'm trying to do my homework here and you're getting in the way!" Jack yelled at the bean (not really expecting it to answer). But when he looked a little closer at his math problem, he noticed that it now said "45x1".

"Wow, I can solve that simple math problem," thought Jack to himself. So he did.

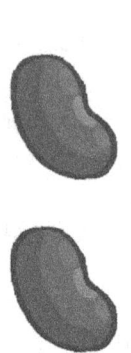

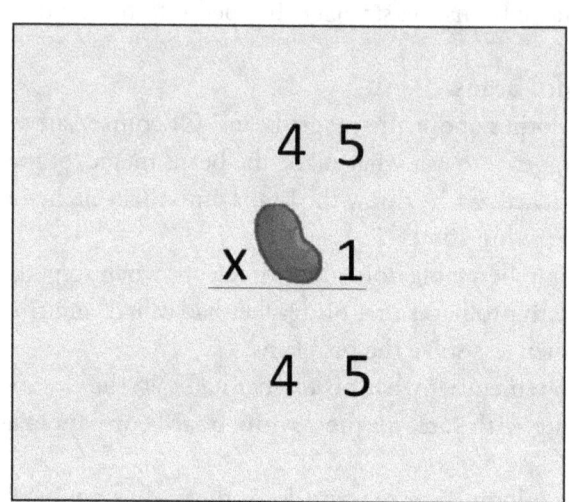

Have students figure out with Jack what 45x1 is, and write the product down. Model what the students and Jack would be writing.

No sooner had Jack written the product of 45x1, than the magic bean jumped over and covered up the "1" in 21. At the same time, a second magic bean jumped onto Jack's page, plopping down right under the product where the one's place is.

Have students make one of their own beans cover up the 1, and place another bean under the product in the one's place.

No matter how hard Jack tried to move the beans, they stayed put. Then Jack noticed that his new math problem was 45x2.

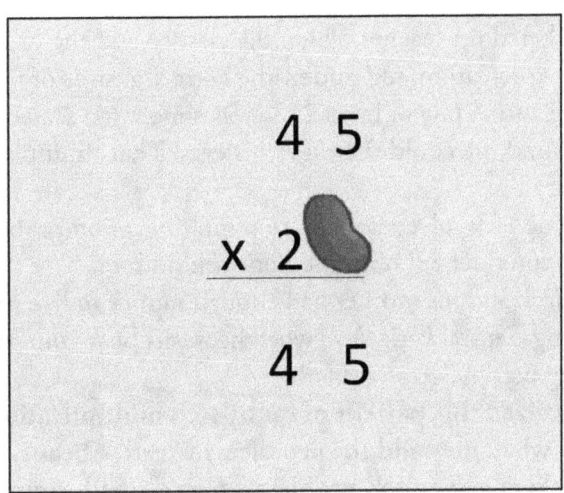

"Hey, I can answer 45x2," Jack thought. So he did.

Model with your students how to solve 45x2, and place the product in the correct place value. Make sure they don't place any digits where the second bean landed.

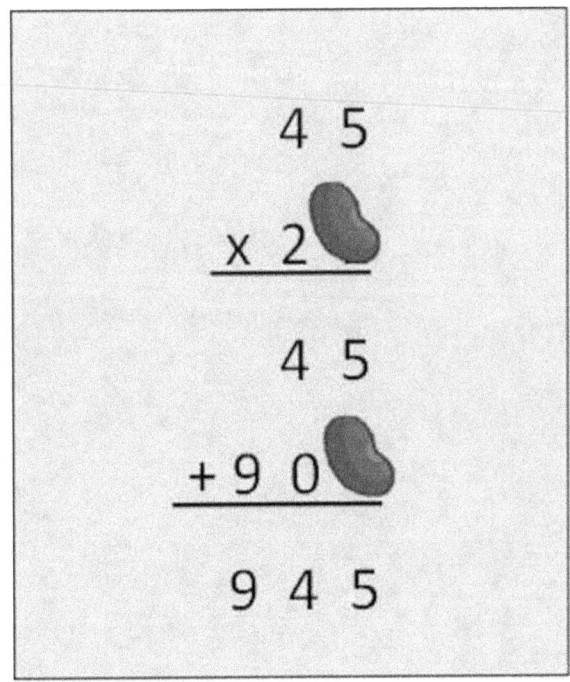

As if by magic, Jack remembered his teacher telling the class to add the products together. And as soon as Jack had completed that simple calculation and gotten the correct answer of 945, he heard a loud thump on the floor behind him. Was it a giant? A pile of bricks? He was almost too scared to look.

But when Jack did turn around, he couldn't believe his eyes! That thumping sound had come from a big bag full of 945 gold coins!

And from then on, every time Jack solved a multi-step multiplication problem correctly, a bag containing the same number of coins as the product fell onto the floor behind him.

So there you have it. Soon Jack and his mother had enough money to live in New York City, in a beautiful penthouse apartment overlooking Central Park. And when he wasn't busy counting all their money, Jack helped all the other kids in the building become great at math!

After students have internalized this pattern of multi-step multiplication, explain the concept of using "0" as a place value holder when they add the products instead of beans. Their brains' pattern-seeking processor will naturally help them solve such problems. Success will come as easily to them as the bags of coins came to Jack.

notes

CHAPTER 3

Measuring Up: How High Can You Fly?

The human brain is a special muscle—it functions as a parallel processor. In other words, it can perform several activities simultaneously. For example, it can process wholes and parts at the same time. The brain also searches for innate meaning through patterning. Any concept related to measurement is therefore a natural fit for the brain. Teachers who fully understand and take advantage of these natural processes can tap into them to help students learn.

Perimeter and Area: Featuring Perry Meter and Auria

Once upon a time a little boy named Perry lived in a beautiful town surrounded by wide open fields filled with beautiful flowers. Perry loved to run around these fields every chance he got.

His best friend, a little girl named Auria, was his exact opposite--maybe that's why they got along so well! Auria loved sitting in the middle of the field, counting flowers that grew in perfect rows.

One day Auria remembered Perry's birthday was coming up, and wondered what gift to buy him. She thought long and hard and visited all the stores, hoping to find something that Perry would love.

The day before Perry's birthday, Auria happened to walk past a store with many wonderful things in the window that caught her eye. She quickly decided that one particular item would be perfect for Perry, and went inside to buy it (she was glad it didn't cost too much).

After school the next day she walked out to her favorite field, carrying the special surprise for her best friend.

Perry arrived, smiling a little because he was pretty sure she had something for him. Auria smiled broadly as she handed him a brightly wrapped box. Opening it excitedly, he discovered the best gift ever--a meter stick!

Produce a meter stick from a drawer or shelf and show it to the class – or better yet, have one wrapped like a birthday gift!

Most of us would think this was a terrible gift! But Perry loved it so much he immediately started using it to measure his favorite fields.

Demonstrate how Perry measured the perimeter of his favorite fields by going around the walls of the classroom with a meter stick, then adding up the length of each side.

Even after he had measured around his favorite fields, there was always something he could measure the outside of. Auria was very pleased with herself.

And that's why whenever we measure around things, we call such a measurement the perimeter, named after Perry Meter himself! (Not really, but it's fun to imagine, isn't it?)

Give students the opportunity to be "Perry Meter" by going around the room with a ruler or meter stick measuring the perimeter of classroom items such as a desk, bookshelf, divider, poster, etc.

Auria's Story

Auria never grew tired of being with Perry as he measured the fields outside their town. She loved sitting in the middle of those fields counting the flowers growing in perfect rows.

These were sunflower fields, so the only thing more beautiful than a single flower was the sight of rows of sunflowers bowing toward each other in the breeze.

After having students show how the flowers might bow towards each other, illustrate a field of flowers in the shape of an array.

Auria was happy counting flowers one at a time. But in really big fields, that took a very long time. "There must be a faster way to find out how many flowers are in each field," she thought to herself.

Once she went to the fields late in the afternoon. The sun was already setting so she only had time to count two rows of flowers. But which rows? She decided to count the row of flowers alongside the path she was on, which had three flowers in it. Then she noticed Perry over at the far side of the field and wanted to say hi to him. As she walked through the rows toward Perry, she counted those flowers, too--six flowers in all between the two sides of the field.

Illustrate the field Auria was walking in, with three flowers in the vertical rows (columns) and six in each horizontal row, making an array of eighteen flowers. This could be done ahead of time on an interactive white board or a poster board.

"Hi, Perry!" Auria exclaimed. "I can only stay for a minute, but I've been wondering how many flowers are in this field. There were three flowers coming up the field and six of them going across the field. There must be a pattern to help me figure out how many flowers are in this field."

Discuss different ways the students think Auria could do this.

"I don't know, Auria, but I'm sure you'll figure it out. You are so smart," Perry said. Then he went back to measuring the perimeter of the field—for the hundredth time!

Auria trudged home thinking long and hard. That night, as she lay in bed drawing, she decided that drawing the field of flowers might help her.

"I counted three flowers going up the field, and six flowers going across," the smart little girl whispered to herself as she drew. Auria continued to fill in each row with six flowers, and each column with three flowers until her picture of the field was complete. She counted each flower and found there were eighteen flowers in all.

"Wait!" she shouted. "There is a pattern! There were three flowers going up the field, six flowers going across the field, and eighteen flowers in all. 3 × 6 = 18. I got it! The number of flowers going up times the number of flowers going across equals the total number of flowers."

Auria couldn't wait until morning, so that she could go back to the field and see if her pattern worked.

Using graph paper, students can create their own rectangular fields and use Auria's pattern to find the area.

And guess what? Auria's new pattern worked perfectly every time. And since then, the inside measurement of a closed shape has been called the "area" after our very intelligent young friend Auria.

Sir Cumference & Diameter

Once upon a time in a faraway kingdom, there lived a knight--but not your ordinary knight. This knight, Sir Cumference, was an adventurer who longed to leave his kingdom and explore the world. His greatest wish was to one day sail all the way around the world.

Draw a large circle to illustrate Sir Cumference's voyage around the world.

After years of planning and preparing, it was time to depart. Sir Cumference and his crew, including his youngest son, set sail. What an adventure it would be! Sir Cumference and his crew saw dolphins jumping and exotic birds flying. They tried to count stars until they ran out of numbers. And they stopped at beautiful islands to play in the warm white sand with sand crabs and coconuts.

But nothing they did or saw seemed to interest Sir Cumference's youngest son. "It's too hot," he complained. "It's too cold. I'm bored. When will we get there?" The only reason Sir Cumference's youngest son wanted to come along was to stop at the first gift shop they came across. Once they had found one, and the boy had his souvenir—a piece of wood carved with a picture of sail boats--he only wanted to go home.

Sir Cumference tried everything to make his son happy, but nothing worked. By the time the ship had sailed halfway around the world, they'd all grown tired of the boy's complaining. "Oh, I'll die if I have to go even one more meter," he kept whining to nobody in particular.

Finally, they'd had enough of the boy's whining and complaining. Stopping at the nearest island with a port, they put him on a ship that was going straight back to their kingdom and Sir Cumference's home.

Illustrate Sir Cumference's path on the circle. At the halfway point, draw a straight line showing the diameter of the circle, representing the son's path back home.

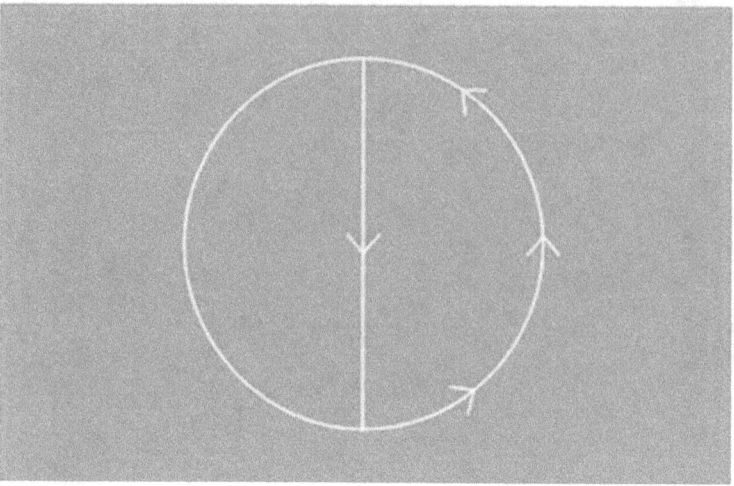

Sir Cumference and his crew continued on their wonderful journey around the world. They had an unforgettable time and made lots of great memories. They had so much fun they kept sailing around the world many times, and never once wondered about the youngest son!

So now, when we measure the distance around a circular object, we call it the "circumference" in honor of Sir Cumference and his love of sailing around the world.

Even his son has gone down in math history. When you cut a circle in half, the line you draw is called the "diameter," since the son whined that he would die if he had to go another meter.

This lesson lends itself perfectly to incorporating kinesthetic movement. Ask students to make large circles in the air, or walk in circles chanting "circumference." Add the short-cut for diameter, while exaggerating the expression, "I will DIE if I have to go another METER! ...diameter." Such a perfect blend of context, concept, facts, and fun add up to the ideal situation to achieve optimal learning.

notes

CHAPTER 4

Geometry: "Gee, I Might Try"

W e've already described the brain as a pattern-seeking processor, always ready to understand communications it considers important. This relates to the old Chinese proverb: "Tell me and I'll forget, show me and I may remember, involve me and I'll understand."

Our interactive activities are specifically designed to actively more fully engage the brain using stimuli such as humor, excitement, anticipation, and contemplation. Students become involved emotionally as well as intellectually. Stimulating the "dance" between context, concept, and fact, we enhance the brain's ability to continually fire these necessary neurons at peak intervals, keeping each student's attention at the optimal learning level. So let's dive into another tale.

Ordered Pair: Starring Grandma and Grandpa Pair

Way out in the country, where "city folk" think people talk funny, lived a lovely old couple named Grandma and Grandpa Pair. This may have not been their real names, but everyone in those parts called them that. That's because they did almost everything in pairs. When shopping, they'd always pick up just a pair of pears, or a pair of bananas. If it didn't come in twos, they just didn't bother with 'em. It was much easier at the clothing store. They could buy a pair of socks, a pair of shoes, and sometimes even a pair of jeans.

As Grandma and Grandpa Pair grew older, they loved to sit in their rocking chairs on the front porch and watch a pair of lovebirds that nested in their tree.

Telling this story with your best "hillbilly" twang will make it even more fun for the students.

Only one thing sort of bothered Grandpa. Everybody knew his wife was a "neat freak." She had to have everything in order, goldarnit! So here's how a typical trying day went for Grandpa Pair.

First thing every mornin', Grandma Pair'd sit up in bed and exclaim, "Grandpa, those covers aren't straight enough. Get up and straighten them." So, Grandpa'd have to slide slowly to the side of the bed and get up to fix the covers.

Students love acting out the part of Grandpa slowly sliding over to the edge of their chairs, then slowly straightening up like an old man.

After making their way to the kitchen, Grandma'd whine, "Grandpa, my toast isn't straight in the toaster. One side will cook darker than the other. Get up and straighten 'em."

So Grandpa'd hafta slowly slide over to the edge of his chair and get up to fix the toast. Perfectly brown toast was always part of their breakfast.

After the dishes had been put away, Grandma and Grandpa hobbled out to the front porch so's they could enjoy the fresh air and gentle breezes. But no sooner had Grandpa gotten comfortable in his rocking chair, that Grandma got started.

"Grandpa, those flower pots must have moved themselves during the night. Get up and put them in a straight line again." Grandpa'd slowly slide over to the edge of his rocking chair and get up to straighten out the row of flower pots (although sometimes, truth be told, he thought of just shovin' them off the railing into the bushes below).

People'd stop by and comment, "Wow, you are such a well ordered pair. How do you do it?" Grandma'd smile and Grandpa'd groan; but neither told anyone their secret.

When dinner time came around, Grandma and Grandpa'd make their way back into their cozy house. "Grandpa," Grandma would moan at him, "Those dishes are not in a neat pile. Go over and straighten them up so they're stacked up nice and tall." So Grandpa'd slowly slide over to the edge of his favorite arm chair and get up so's he could straighten up that wobbly stack of dishes.

Most evenings, Grandma loved to sit by the fireplace and read while Grandpa got to whittlin'. But no sooner had she and Grandpa gotten themselves comfortable when Grandma would snort, "Grandpa, I can't reach my lovely stack of books. Bring them to me." So, Grandpa'd slide over to the edge of the couch and get up to fetch Grandma's stack of books from the shelf.

Every day was pretty much like the day before. People would ask how they stayed together and kept things so ordered. But no one was able to get their secret out of them.

Years and years went by, and Grandma and Grandpa passed on—on the same day! One day, their great-great-great granddaughter was exploring the old house. Near the fireplace she discovered a loose board. She pried it up to see what was underneath. That's how she found this old dirty piece of paper with writing on it.

She excitedly opened the paper and read, "Our Secret To Staying Ordered" by Grandpa Pair. "Well, here is our secret. Whenever Grandma Pair ordered me to do something, I would always slide over to the edge of my chair and then stand straight up. She never had to ask me twice."

The secret was finally out. We all love things to be in order; so in memory of Grandma and Grandpa Pair, whenever two numbers are together in parenthesis, we call them "ordered pairs."

And when we need to graph such a pair of numbers, we always slowly slide across the graph using the first number, and then go straight up using the second number.

As you demonstrate how ordered pairs work on a graph, encourage students to s-l-o-w-l-y slide to the edge of their seat and then stand straight up.

As with the previous lesson, you've just actively involved and engaged students in the lesson. Not just using pencil and paper gives you many opportunities to keep their brain neurons firing. Before students can learn, they have to pay attention. That is the key to success!

notes

CHAPTER 5

Data Analysis: Show Me the Data!

As discussed earlier, successful learning requires focused attention. But it also involves peripheral perception, learning that takes place unconsciously. As we learn, our brain stores this information in two types of memory banks – spatial and rote. It understands best when facts are embedded in natural, spatial memory. This memory is formed after the brain gathers sensory information (sights, sounds, tastes, touch, etc.) about its surroundings. Then it records information about the environment and spatial orientation--"seeing the big picture." Learning is enhanced by challenge, by authentic experiences, and by active processing. On the other hand it is inhibited by threat, embarrassment, or guilt.

Mean, Median, Mode and Range: Down on the Farm Style

Way out in the country, where some city folk think people talk funny--you know what I mean--there lived a very small family. I say small 'cuz it was jest a Ma and a Pa. They wanted children real' bad. Day after day Ma would lean over to Pa and say, "Sure wish we had some children, Pa. 'Specially some strong, strappin' boys to help out with all the work here on the farm."

Best to bring out that "hillbilly" twang again to tell this very humorous tale!

One bright mornin' as Ma and Pa were reading the newspaper, Ma noticed an ad in the paper. It was about a poor strong, strappin' boy that had lost his Ma and Pa (naw really, he jest couldn't find them anywhere! He even looked under the bed and between the cushions on the couch!) and was looking fer some new ones!

Ma and Pa jumped into their pickup and rushed into town to check things out. Soon as they saw that boy, they knew he should be part of their family. So they took him home.

Ma was so happy! "Pa, now I have my strong, strappin' boy!" she exclaimed. "Boy," she turned to the boy, "what's yer name?"

"It's Mean," said the boy. But, he really wasn't mean - he was jest average.

Pause here to explain to the students that in math, "mean" means "average".

Well, Ma and Pa and Mean (who wasn't really mean – he was jest average) got along perfectly! They would milk the cows ever' morning and feed the chickens ever' night. They would till the soil and weed the garden.

Explain that since Mean is the oldest boy in the story, he has to do more work than the brothers who'll be coming along in good time. And that in math, finding the mean takes the most work, but still not too much. Show students how to find the mean (or average) by adding up a set of data and dividing by the number of data you have. For example, the average for the set of data 3+6+9 would be 18÷3=6. The mean, or average for this data string is 6.

But before long, Ma got a hankerin' fer another strappin' boy to have aroun' the house so that Mean (who really wasn't mean - he was jest average) could have a younger brother to horse around with.

Continually repeating the phrase 'he wasn't mean-he was jest average' drives home that these two words are synonyms. Urge students to begin chanting this with you.

And, sure enough, jest a short time later, in that newspaper Ma and Pa read each mornin' came another ad fer a Ma and Pa who might be lookin' fer a young strappin' boy. Ma and Pa rushed down to the town's general store and picked that boy right up to take home with them.

Now, this new boy was a bit younger than Mean (who wasn't mean – he was jest average). His name was Median. Ma and Pa thought that was a great name 'cuz Median sounds like medium, which is the middle size.

Show students with a set of data how to find the median number (or middle number) by arranging the data in numerical order and systematically finding the middle number. For example, for the set of data 12, 27, 32, 54, 98, the median number is 32. It is the middle number in the set. When finding the median number, make sure to point out the set of data has to be in numerical order from least to greatest or greatest to least.

Mean and Median got along jest like any two brothers would. They worked hard and played hard with each other ever' day.

But, it wasn't too long before Ma said to Pa, "Pa, it seems to me that we need jest one more strong, strappin' boy to round out our family." And sure enough, it wasn't much longer before she got her wish. In that very same newspaper, another ad appeared. This time there was an adorable little boy who needed a family to take care of him. And somehow Pa jest knew they would be addin' to the family again that very afternoon.

When Ma and Pa made it to town, they jest fell in love with this young 'un. They loaded him up in the car and took him right home. Oh, how they all loved this little boy named Mode. They loved him so much that they gave him everything he ever wanted. Oh, was he spoiled! He got the most out of all the family members. But that's ok, 'cuz Mode means the most!

Point out to students with a set of data such as 32, 54, 62, 23, 54, 71, that the mode is the number that appears the most. In this case, it would be 54.

Well, you would have thought this family was complete. And fact is they did, too. Until one day the three boys, Mean, (who wasn't really mean – he was jest average), Median, (which sounds like medium or the middle size), and Mode, (which means the most) all came to Ma and Pa with big pleading eyes and said, "Ma and Pa, could we get us a dawg to play with? PLEASE?"

How could they say no, especially to little Mode? So, the boys hopped into the back of the pickup for the drive into town, and they went lookin' fer jest the right dawg. And sure enough, there on the side of the road was the mangiest mutt you ever did see. Well, Ma took one look and knew that dawg was fer them. So, Pa stopped the car and in the dawg hopped.

He must have been a stray, 'cuz he didn't have a collar on him or anything. Oh, those three boys were happy! As soon as they got home, out jumped the dawg, follered by the three strappin' boys. They all raced over to the pasture and began to play.

Now, this dog loved to run! In fact, his favorite game was to skedaddle from one side of the pasture to the other, where he'd stop at the boys' big feet and look up at them, tail waggin', tongue hangin' and he'd seem to say, "Are you proud of me? Hhh, hhh, hhh?"

The students get a kick out of the teacher racing from one end of the room to the other just like the dog, illustrating the concept of range.

That's when the boys come up with the greatest name fer this here dawg. They named him Range, 'cuz he felt right at home, home on the range; and (by the way) when you find the range in math, you see how far it is from the biggest number to the smallest.

Using a set of data such as 14, 36, 52, 64, show students how to find the range by subtracting the smallest number, 14, from the largest number, 64 to get 50. The range of this data is 50.

As Ma and Pa looked on, watching their three strong, strappin' boys playin' with that mangy, wonderful dog, they knew their family was now complete.

Using these, and other stories you as the teacher make up, you will begin to see three things happening.

First, students will start smiling whenever they do math. They will look forward to math with excitement and eagerness.

Second, you will start to have many more of those "ah-ha" moments as you see students making connections in their brains, understanding concepts, and using the skills needed as building blocks for more difficult mathematical concepts.

And third, the retention rate will exceed both your and your students' expectations.

Since incorporating brain-based learning into my class room, not only have I enjoyed teaching more, but my students enjoy learning more, and their test scores have skyrocketed!

I hope and trust you will have the same results. As the saying sort of goes, "Tell me, and I will forget, show me and I will remember some, involve me and I will understand."

My best wishes to you and your students as you continue to change lives every day working in the greatest profession on earth!

notes

Author Biography

Shauna M. Johnson, M.Ed, NBCT, is an elementary education teacher who earned her BA in elementary education from Weber State College in Ogden, Utah. Her passion for learning propelled her to obtain a master's in education, curriculum, and instruction at California State University, Bakersfield.

Johnson earned her National Board Certified Teacher status (NBCT) and has taught preschool through fifth grade over her twenty-five years in education. She has worked with students in Texas, Minnesota, Utah, California, and Virginia.

She's focused her career on creating optimal learning environments and curriculum for schools and school systems. She's worked as a mentor teacher, served on leadership teams, facilitated change in underperforming schools and helped write and implement new curriculum in mathematics, science, social studies, language arts, and writing.

Johnson now lives in Atlanta, Georgia, with her amazing, supportive husband of twenty-five years. They have three grown daughters and one adult son.